# PS: See You Later

## Taking the Fear out of Death

Thousands of people in the entire world are so afraid of death. But what is behind the fear of death and how does death actually work? One thing I will tell you now is that the fear of death isn't literally the fear 'OF IT'. We will cover all these questions above and provide answers in this book. I am going to share my experiences with death, both in animal and human formats. These will hopefully be able to not only show you that you are not alone, but also be able to show you death in a different light.

As an intro, I want to let you in on a fact about death. More than 7 billion people live in the world today and that number is constantly growing due to births. It has been recorded that the birth rate is overpowering the death rates. So, for every death that occurs it is estimated that there are at least 3 times that amount of babies born at the same time.

I will now go into the larger portions of this book discussion various experiences of 'death'.

# My Human Experiences of Death

I have had quite a few experiences of people that I knew very close and those who I knew fairly close that have died. I do not use the word passed away because I want to live in the reality of the word. I am not replacing the word death with an antonym because death is death. It can be referred to as other words, which is fine, but sometimes people find it easier to use other terms rather than the actual formal term. If you want to overcome your fear of death, then you need to not avoid the word. You need to embrace the word and I will also again explain that later.

I find it very helpful to walk through all the parts of death that it holds. I also find it extremely helpful to write about my feelings and to write letters to the person I love that has died. It makes it more personable so that I am not pushing away the person or forgetting to include their spirit in what is still going on for me here on earth. One of my most favorite terminologies that I use for describing death is this: Every state and country is like a room in a huge mansion. We travel through each of these rooms for however long we do. When we face death, we are just going to the entrance of the mansion. It is like we are taking a vacation and some of us leave vacation before others do. Vacations are fun, stressful, emotional (good and not so good), exciting, interesting, etc. Every emotion and feeling is incorporated into each and every vacation a person takes on earth. I feel that we are on a vacation and when we die we are returning home or returning to the front of the mansion that resides. We are here to work, play, make healthy arguments, love, give, get and create for our God (I do not mean to offend anyone if you do not believe; I have a section for you all soon to make this concept easier to accept even if you currently do not believe in Jesus Christ).

Here are some of the writings that I have written when I have had someone close to my die. They are as follows underneath this paragraph.

My grandpa was a Korean War Veteran. He fought in the Korean War and survived. While he was there, part of his job was to operate the tanks, which are huge. He had pictures of him and his friends from the war sitting on the rows of these huge tanks. He didn't talk very much about the war but I know he always remembered it very well and when I would have a question at times, he wouldn't hesitate to answer it. My grandpa after the war worked for the Augsburg Publishing

Company and he took his job very seriously and performed it beyond the standards. I have to thank him for encouraging work ethic to all of us. He would take the train everyday to and from work. He fought against the arthritis pain and that has helped me deal with my aching, as I remember that he did not give up. My grandpa was also a father, a husband, a brother and a very caring grandpa. He was my dad's father and my aunt's father. I know that anyone can be a father, but it takes someone very special to be a dad. He was definitely a dad, as he was very special. He took it beyond the word and made it mean what Jesus Christ wants it to mean. He also was a fantastic husband. He reminded all of us that love is strong between a husband and wife… that it doesn't have the chance to break. This is because you need to remember the special day you asked the woman you loved to marry you and the day that you spoke your vows, and not just speak them to an engaged audience or each other as a custom, but in front of God Himself. He showed what it was like to be a supportive husband and a person that would give and take to keep a relationship strong. I only hope that if some day I am in a relationship, I will follow the ways he was a husband. My grandfather was also a brother to a very special family. My grandpa would talk to me about his brothers and family that were either gone or still here. When I started playing the violin he talked about his dad playing the violin and said his brother did as well. Well, his brother says he didn't play it that much... and he laughed at that after the service. I will never stop playing the violin; it is now officially a family tradition! Family was so very important to my grandpa. My grandpa was the best grandpa, veteran, husband, father and brother. He will also always be a friend to many and to those who always look up to him.

# What Does Life Teach Us?

## What my grandpa taught me

My grandpa taught me SO many things and I will never forget all the important lessons about life that he taught. My grandpa had this United States glass tray sitting in his room where his desk was. I asked him one day why and where he got it from, in which he answered "Read the words right there..." And I did. He said, "It represents freedom." My grandpa was very much in to reading articles on who was going to be promoted to governor and president, and even who was going to be promoted or demoted at stores in the state of Minnesota. He had high opinions on everything. He had vast opinions on the war that is currently going on today in Iraq and told me several of his feelings. He did articulate that our world is trying so hard to stay free and that freedom is never really free. He told me that no matter how strong our opinions are, that we need to respect the one who makes the decisions because he or she helps keep us safe and keep the world in working order. He asked me, "How hard would it be for you to control the whole world? If you were the president or mayor, you would think differently very quickly." My grandpa talked about his uniform that he wore in several of the pictures he had and that he talked about. I asked him if he still had it and he said, "Oh yes… somewhere dear." I still to this day have never seen the uniform he wore in the war and if I ever got a chance to, would make that moment last. My grandpa was always about making moments last and be full of meaning, even in the smallest of them. He had the funniest and most bright laugh I ever heard. It is in my mind today and every time I am really sad, I think of his laugh and it just makes me giggle and smile. My grandpa also had such a kind and warm touch. When he would give me a hug, he would be gentle and yet he would give you a hug that felt like he was saying inside "I love you so much and don't you ever forget it." My grandfather would call me on the phone several times and I would love chatting with him. The thing he loved talking about the most was the weather! He would turn on the TV and get out the newspaper clippings. He would talk about how the precipitation was where he was, how the snow was falling outside the window into the planters that were just planted (and he'd laugh! – sorry grandma), and talk about the hosts on the weather channel. He would even correct their mistakes. It was interesting that the last chapter he was reading of my book that I sent to him after writing was on weather. My grandfather also talked to me about my book and would ask me questions about it and myself. He would go through stories with me and talk about little things he remembered that related to what he was reading in the book at the time. It made me feel really good to know that my grandpa cared so much about what I had written and would love to talk about it with me. One time he called, I had just woken up with the stomach flu. I had told him that my mom was going to take me to the hospital after she got home. He said, "You make sure you go and get fluids so you get all better. Take it easy and rest lots. You make sure they take care of you there." I asked him to not tell anyone else because I didn't want them worrying and he reassured me that it would be just between me and him. Any time that I asked my grandpa to pray for me he told me that he always did and was already on it! Sometimes I would get embarrassed by things that were happening to me and I would say, "Grandpa, can I ask you a question?" He always took the time to listen and give me such great valuable answers that I think of today. I am reminded that if I am to pray to Jesus, that I pray with faith and just ask Him, and ask if He was standing right next to me (He is). But ask as if I saw him with my eyes; I guess this is the way to develop your faith. To lighten up lots of situations, he would give the salute sign and give a big smile! If I were to feel worried about something, his question would be, "What for? Just let it happen and move on. You are going to be ok." One thing in particular that I remember was when we were going to walk around and look at Christmas lights. My grandpa, because it was cold and it was hard for him to walk with his arthritis that night, said he would just have a nice snooze in the car until we got back. Well, that snooze didn't last very long at first, because he accidentally set off the car alarm! And, funniest part is that we couldn't figure out how to turn it off, well at least for a minute at least. My grandpa had such a great attitude about things and would give these really simple answers that he would kind of make into a joke. I had asked my grandpa one time, "How long until you need to fill that up with air again?" He responded with a laugh, "It waits until we are ready to leave!" My grandpa had a lot of severe arthritis but he did his best at making things work and living his life in the best way possible. My grandpa would tell me that we need to not waste time complaining about any things, but doing something about them because talking only gets things so far. He said that it was ok to cry and that all men cry, but they just try to hide it. When he was with me, he hadn't been sure if he would be able to try the wii bowling with me. He stood up, put it in his hand and bowled a strike… on the first try! He said, "Well, I guess I can beat you, ha?!" I laughed and he played a whole game with me and of course beat me with ease. He then sat back down to rest and being himself, just had to add a thank you. He thanked me for showing him the wii bowling and said that he had a blast playing it with me… he was glad that he tried and was able to do it. We would sit and he would try to teach me how to play poker on his handheld games which I loved sitting on the couches and doing with him. I would nod my head and the whole time didn't understand a word! But that didn't matter to me. It was wonderful hearing my grandpa having so much fun trying to explain how to play this very complicated game. And his high scores would keep getting higher too almost every time he played. My grandpa also taught me tricks of the trade when it came to playing Solitaire. I finally won a game when he showed me how to put cards back up and down from the Aces pile to fit more cards in from the deck. I now play solitaire to this day and love winning, all because of my grandpa's teaching! When I win, I look to him and say, "We did it again gramps!" In turn I tried to teach him Sudoku since he loved numbers so much. When my family went out and I decided to stay home with him, we went through and played a few games together. He talked through it with me and completed a whole one almost completely by himself! He told me I was a really good teacher. And then he also told me he could never handle playing it on his own though! Sometimes, when I was feeling sick myself, I didn't want to participate in some activities. But, I knew that my grandpa sometimes also didn't want to either. So, he stayed with me. I had some movies that others in the family were not interested in even viewing once. When they left, my grandpa told me to put the DVD in. He was so interested in the movie and at the end said it was his favorite movie of all time! See, when you are as open minded as my grandpa was, you may come to love things you thought looked not so neat. You have to give them a chance, which is what my grandpa also taught me. My grandpa talked to me about the different fruits and vegetables they used to grow in the garden they used to have. He and my grandmother tried to talk me into loving these horrible dried apricots.  It did not work and when my grandma went to take them out, he'd tap me on the arm and say, "Guess what your grandma is taking out?  Your favorite!"  And he'd laugh as my face turned into a look of horror.

On a more serious subject, my grandpa was the one who taught me several things about my dad and his sister, and my grandmother.  Sometimes when I talked to him, he would ask me to keep my dad in check and make sure he was fixing things around the house and eating healthy.  He told me that him and my aunt were close and he was very lucky to have them as his children.  He was so proud of my aunty and my dad for raising families and being there.  He also told me how much my grandmother does for everyone and at her job.  He said, "Your grandmother works so hard for everyone… And she loves everyone."  I had talked to my grandpa about my cousin's wedding that was coming up and on some occasions before that.  I had told my grandpa that he and my grandmother had been married for so long and I wish I had some advice to give my cousin and her soon to be husband.  My grandfather taught me about holding on to one another when times are tough and that

that is what makes you stronger and love each other even more.  He said that my grandmother was an amazing mom and wife and would always be his beautiful bride.  He said that a couple needs to hash things out sometimes so they don't keep it inside and said that you have to have jokes inside a marriage.  But most of all, you need to remember the time when you said "I do" and it will get you through anything as a couple.  He said, if you remember that moment, then you can get through anything that happens.

Can you who are reading this, see all the different types of things that life teaches us?

I will know go onto the chapter about how it feels to say goodbye to someone who is about to die. For many people this process is very scary and for me it was at first but then I realized why NOT to be afraid.

## Saying "See You Later"

The day came where we had to get on the phone to talk to my grandpa for the last time here on earth. I was shaking and so scared. I thought I could handle it, but I realized what was happening; I was going to talk and hear my grandpa's voice for the last time until we meet again in Heaven. It made me so sad because I wished I would be able to be there so he could see me and give me another hug. I selfishly wanted him to stay but at the same time I felt a peace in me that I haven't felt before. It was like Jesus was there and telling me, "He is READY…" I thought about all the different things in life he'd done and that I had been there for. It was really amazing and quite simple; He had lived life so fully and had an amazing amount of years experiencing life on earth. He was going to go to Heaven where it is perfect and joyous, and that was ok with me. The reason I cried so hard when I went to say that I love him was because I was afraid that I would scare him. I was afraid that he would figure out what was really going on if he hadn't already and that my emotions would scare him. The last thing I wanted to do was scare him but then I felt so much better when I told him and was told I would be ok. I knew then that my grandpa was not afraid. He was just going to Heaven and would 'see me later' as we would always say to each other. With him and I, the attitude is NEVER goodbye, but rather "See you later". And that is so true; thank you Jesus Christ for making that possible and a reality. So again, I now say without fear or sadness, "See you later."

An extremely important part of the death process is the recognition, the funeral and yes, the burial. This can be such a difficult time for people because they realize that this is the last time for a long time that they are going to be able to touch and see their loved ones. Many people wear the color black during this time especially to show that they are in mourning, but for me, I think that trend should be changed. The reason being is that we should not be afraid of death and feel completely doomed that we are going to have to wait a while to see that person again. This person is now not alive anymore in their physical body on earth (on vacation). They are not crying or sad... so why should we be wearing a color that represents such darkness when death really is all about light and a fresh beautiful start to a wonderful life thereafter?

## Why Do People Fear Death?

In the very beginning of this book I said that I would talk about why people fear death. I also said that one thing I will tell you now is that the fear of death isn't literally the fear 'OF IT'.

The reason people REALLY fear death:

1. I will leave people behind and they will suffer

2. It will be so painful

3. I will give up all control... I am NOT in control anymore

I want to talk about how we worry about leaving people behind. What is so interesting about death is that if a person is told they are going to die in two months, they are more scared for the people who will be left behind rather than initially being terrified of the fact that they are going to die. This has been a census of those who were young to old in age in this scenario. You need to understand that people die one day and you will be included in that. Those around you will mourn, but they will survive... they WILL survive. And hopefully by reading this book, they will not be so intimidated by death and will look at it as the fresh and pure start that it really is.

The major issue of many people is that the death will be extremely painful. They fear the most that they will feel things shutting down inside them and that their breathing will feel scary… that the breathing will cause them to feel like a fish out of water.

I have to tell you that I have had three experiences of not being able to breathe. One is when something was stuck in my throat, the second was when my trachea cap collapsed under water and the third was a mini asthma attack where I was given steroids.

Having an asthma attack is where your throat and vessels are literally closing up. It is a very scary feeling. However, what people do not understand is that when you die, your throat does not close up but instead just stops working.

The times where my throat was closing up was very scary and I got hot. However, when it happened where my trachea cap collapsed under water (which means it just stopped working) I felt like I could see really bright lights. I got very tired and I felt like I was falling asleep into a dream. In all honesty, it was not a scary feeling. When we die, our throats act like that collapsed trachea thing that happened to me underwater where my breathing just stopped working. My throat was not swelling up and that is the huge myth about death and breathing. This is what makes people the most scared about death when it comes to terms of feeling a very horrifying experience. This type of breathing abnormality when dying causes extreme sleepiness and usually causes the person to fall asleep. When I had this experience under water and coming out of it that is exactly what it felt like. Believe it or not it was not scary, but rather just a feeling of being really tired and falling asleep into bed when you can't keep your eyes open any longer after a really long day and night. When there is total unconsciousness, the body is asleep and protecting itself from knowing the rest of the body is dying. The body does start to shut down and every time it does the body just gets more tired and sleeping occurs more. Sometimes the body will twitch, but this is not to be confused with a person feeling turmoil while that is happening. An analogy that may help you understand is this: A lizard is dead, yet its tail or legs continue to move. The parts in your brain that control movement are firing... that is IT. This also goes for sounds like whimpering, etc. People who are dying will experience a type of feeling of disorientation and may say things they don't know what they are saying. The doctors will tell you that they have no clue what is going on. I had taken a medication that would be similar to what disorientation would be like and amnesia, or a total blank in understanding. I was told 4 days later that I ate, slept, took a bath, etc. with help. I have NO recollection of this. This symptom is actually a blessing in disguise. People should not fear the symptoms of death because it has been polled, seen and I have experienced near experiences as far as it comes to the breathing aspect. The throat does not close… it rather stops working. Do not fear that you will be in severe agony or that your loved one will. Death is falling asleep and moving from your vacation to your totally pure and joyous mansion in Heaven.

The third concern is that you are not going to be in control anymore. As long as you have breathe at all, you have some control. It is human nature to want to be in control and the thought of not being in control is so frightening. But let me reassure you that what you have just been told about death, not to fear losing control. Your brain is controlling your knowledge that you are dying and is able to numb what would not be handled. It is working for you in your favor, if you can understand what I mean. It is very powerful. Do NOT be afraid of death, but instead embrace it and learn more of it. You will then eventually feel better and be able to live the life you are living greater and larger than you can ever imagine for however long that may be.

## Decisions- Cremation vs. Embalming

The caretaker, I, along with my whole family gave us time to think about the hardest thoughts beyond the hardest time. Upmost, they gave me the decision since she was mine (animal death; same as human death on a different level) and we were connected in such a deep level. Most vets in many of their clinics take your loved one who has now passed on and cremate them. The thought to me of having that happen to her beautiful body and eyes, whether human or animal just seemed so wrong. I felt as if she wasn't getting any respect. I wanted her to have a type of handling like other humans can ask for beforehand. I wanted to be able to visit her everyday and know where she was, know that she was still covered in a beautiful blanket. It was ripping me up inside just to think about it, so the caretaker told me there is the other way of going about it, but most people do it the other way- handle them with cremation. I felt so sick inside because to me she didn't deserve that, but that she deserved something beautiful.

What I have to mention here is that it is suggested that people meet with a local funeral director before their body will die and go to Heaven. It makes a person feel more comfortable if they know who will be handling their 'shell' after they die and go home. It is encouraged and taking the time to bond with the person can give such peace.

I will now continue on with what happened in the above scenario.

What I did not think about until later, after I had said for them to go ahead with it, was that I think about the psalm songs of the bible and the verses in the bible. For instance, "From dust we came and to dust shall we return." That makes sense when you read how the first beings on earth were created and how you think of natural life, again the wild… how when we walk on the very soil of everything, every layer has a story and that may be as special as if it had just been placed in silk in a box. We get to excavate these days and we get to see how we have in a way transformed. We can appreciate our lives but we also do so with humility but not shame. This is of course not the same ordeal, but I see it as a sort of an analogy. Maybe the scriptures and history have come together in harmony and shown us really that our world is a circle of coming and going, but going back to where you came from; your original origin, that of Jesus when he created Adam and Eve.

Over time, the other way of handling them would turn out similarly. Still to me I wanted her sweet and caring body to be wrapped in silk, which brings me to my next reminiscence, the psalms and songs. In many of the songs it talks about, "I will go, go, go into the refinery and come out clothed in silk." I realized that Jesus had already taken her real life, her spirit with Him and that cremation meant similar to that of what the psalms and songs were saying. She was going to come out clothed in silk, the silk of Jesus.

It wasn't going to be a negative or put down on her beautiful face. She was coming out better, only in a new way. It may be hard to think of that as you think that what is in a small white urn is anything close to silk, but it is because human intelligence I believe gets in the way.

This is yet another example that I am going to try and help you understand. Lava seems to destroy things, it is like a refinery, but what comes out of it is beautifully colored stones and pebbles. Everyone who can get their hands on these miracle stones try and get them because they are so valuable, so precious. The lava acts as a refinery and so does the cremation. She is just as beautiful, if not more... only in a different way.

Different is a good thing, although it may be hard. There are pet sanctuaries and sites that you can either clothe your pet in cloth and bury them or refine them in an urn there as well. It is just like the decision you make for a human family member or friend. You can talk to them every time you visit, but you would have to go there. Since this experience was with the death of a pet I chose to have her new beautiful face and body clothed by

Jesus' refinery in my bedroom on the desk just away from my bed. I always feel so much closer to her in a physical sense still. I say goodnight to her and I tell her how much I love her. I put her pony she loved by it and pictures/blankets surrounding it. I believe it is showing great respect in letting her Fly Away Home. She is Home now in Heaven but refined in love... and in my home so that I can feel closer until I meet her again someday when she comes running to me. All the other people and animals in heaven will be asking, "Why are you running away?" And she will say, "To come jump into the arms of my mother..." For my grandpa, he was buried at a cemetery. The same feelings and thoughts resonated through my mind on all of what I wrote about above. I will describe more now about the importance of services after someone dies. Also, an extremely important part of the death process is the recognition, the funeral and yes, the burial. This can be such a difficult time for people because they realize that this is the last time for a long time that they are going to be able to touch and see their loved ones. Many people wear the color black during this time especially to show that they are in mourning, but for me, I think that trend should be changed. The reason being is that we should not be afraid of death and feel completely doomed that we are going to have to wait a while to see that person again. This person is now not alive anymore in their physical body on earth (on vacation). They are not crying or sad... so why should we be wearing a color that represents such darkness when death really is all about light and a fresh beautiful start to a wonderful life thereafter?

# The Service

My grandpa was honored at Fort Snelling. The honorable 21 gun salute and music was beautiful. I jumped about 5 inches off the ground it seemed each time they fired the guns. Each time they did it looked as it did in the movies but this time it was for real. It looked like fire going through the air. Up in the air, the smoke made the symbol of the cross. I think that was my grandpa and Jesus saying they thought the remembrance was beautiful and that was their way of letting us know it was ok. I was feeling very sad because I was missing my grandfather not standing next to me anymore but I was feeling something else that I just can never explain. I felt almost proud or something. I had and have so much respect for my grandpa and what he stood for. My grandpa stood for world freedom, what was right, strength, unconditional love, laughter and joy. When they were folding the flag and handing it to my grandmother and shaking our hands in their uniforms, I felt as if my grandpa was looking down and smiling. I felt as if he was shouting, "God Bless America." Everyone after this salute and ceremony was visiting. Everyone was getting ready to leave because there was going to be another memorial for a soldier soon. I said, "Are we not allowed to say goodbye?" I thought that maybe this was a military ritual or maybe something that people did at the fort to be respectful for the soldiers. I went up to where my grandpa was resting in peace because I felt like I never completely got to hug him goodbye. I wanted to hug my grandpa one more time before leaving him. The first two times that I hugged and kissed him in peace, I felt like I could literally feel my grandpa's gentle, yet firm hug around me. I felt like I was feeling him hugging and kissing me back. And the third time and kissed him and hugged him in peace I said, "Ok grandpa... You are no longer physically here. Please go back up to heaven and know how much you are loved. I will hug and kiss you again in spirit, but not physically again until I see you in heaven. But that is ok, because now you know nothing but happiness... and I'm ok with this. You are really home now and I'm just hugging and kissing in your honor now." And with that I felt a strange feeling of peace as I looked up at the sky and went on to be driven back with everyone to the church. I know my grandpa was saying, "This was beautiful and now it is time to move on." He was also saying, as we decided to say to each other each time we had to leave, "It is NOT goodbye, it is See You Later." So, I said "I love you, and see you later!"

I want to quickly mention here that people can become very nervous around head stones. A head stone should not be looked at as something scary but instead something of love. It shows the world the beauty and years of life they lived. It is there so that future generations will never forget your loved one who has died. It is an honor.

In this next chapter we are going to talk about the importance of writing and keeping the person who is physically dead into everyday life as respect as well as for joyous purposes.

# The Letter

It is important to keep on living the spirit of the person who is dead. It helps you and it shows them respect. I find it helpful to sometimes write letters to the person who has died to show that you can still talk with them although you may temporarily not be able to hear what they say back.

Grandpa,

Thank you so much for everything you stood for, who you were and being such a wonderful and caring grandpa. You put a whole different perspective into the word grandpa, dad, brother and husband. I now understand fully the gravity of those titles. Thank you greatly for being such a great grandpa and all the wonderful lessons you passed on to me. Thank you so much for your service, your gratitude towards this earth and all the people in it. Thank you for your love and joy that you spread to everyone; as you say "Smile big everyday!" Thank you for just being you.

I love you always and will never forget you.

PS: Until I see you again, "See you later." Love, Kristina

Do not feel silly writing these... this is a part of honor and not forgetting still inclusion.

# Songs of Death

I think that songs playing when someone dies and those around experience the death before them, music is so important to give words to each part of the process of dying; the living, getting ready, dying, and finally getting a fresh start. I hear many times of people who put on such sad songs when a person dies and that only exacerbates the fear as well as the pain the others are feeling who have not yet died and passed on through from the vacation to the mansion. I instead suggest different songs. Here are some songs that I believe coincide with the different stages of death that will not cause pain but instead cause a sense of relief and a lesson for all of us who have not experienced it yet.

Grandpa,

There are a few songs that I think would describe your attitude towards life here on this earth and transitioning to life in Heaven.

Getting Ready:

~I'm going home to the place where I belong

~Where your love is always been enough for me

~Don't regret this life that was chosen for me

~These places are getting old, so I'm going home

~I'm going home to the place where I belong

~I'm going home . . .


Getting There:

~I can only imagine what it will be like when I walk by your side

~I can only imagine what my eyes will see when your face is before me

~I can only imagine

~Surrounded by your glory what will my heart feel? Will I dance for Jesus or will in all of you be still?

~Will I stand in your presence or to my knees will I fall?

~Will I sing hallelujah will I be able to speak at all?

~I can only imagine . . .

~And when all I will do is forever worship you!

The 2 Songs That Could Define Your Look @ Everything:

(A song of wisdom and positive attitude)

~When I find myself in times of trouble Mother Mary comes to me, speaking words of wisdom Let It Be

~And in my hour of darkness she is standing right in front of me, speaking words of wisdom singing Let It Be

~Let It Be, Let It Be, Let It Be, Let It Be

~ Whisper words of wisdom Let It Be

~There will be an answer

~There is still a light that shines on me

~I wake up to the sounds of music

~There will be an answer Let It Be


(A song of pure peace and extravagant joy)

~Some glad morning when this life is over I'll fly away

~To a home on God's celestial shore, I'll Fly Away

~I'll fly away old glory

~When I die hallelujah by and by, I'll Fly Away

~Like a bird from prison bars has flown, I'll Fly Away

~I'll fly away old glory

~To a land where joy will never end

~When I die hallelujah by and by, I'll Fly Away!

See? Death really isn't so scary… It is God's party to us as we are departing earth and coming to our real home, ha?!

I know that all of those who have died are walking strong and praising God… They are SO much in peace and joy; A kind of peace and joy that can't even be thought of. For God tells us that… no ear has heard, no eyes have seen, and no mind has understood what God has planned for those who love Him. I am so glad that Jesus died so that we will be together forever, worshiping Jesus and in His pure white presence with a world that is completely in harmony. How wonderful is that? How wonderful is it that death really means

harmony, joy and everlasting peace?  Death is not
scary and depressing in its original sense... it can
be difficult but should not be scary and depressing.
It should be the opposite of that in its own way.

## For Those Who Do Not Yet Believe

These are my couple questions for you as well as my advice and pray along prayer...

1- If you have one option to take a chance on either NEVER seeing those you love again or having the chance to live again in complete harmony with those you love, which would you choose?

2- Should it be that hard to take a chance? What could it hurt? It cannot hurt anything but instead make everything better. When you die you will either know or not know. So, if given the chance to accept God into your Heart and Spirit, will you in order to see your family again take a chance that you would never regret?

3- If so, pray this with me... "Dear Lord, I am unsure at this time if you really do exist. It is hard to feel that you really do exist because I cannot with my eyes see you and touch you with my hands as other people are in the flesh. However, I also know that I cannot see the wind or the ions in it, yet I believe and know that it exists. Also, I cannot prove that you do NOT exist. I want to take a chance on seeing those I love again in a place that is only in perfect harmony at all times. I want you to understand how I am having a difficult time. I want you to help show me through your different signs that you indeed do exist. I want to take that chance of saying "YES I ask and accept you into my Heart and into my Soul." If this is true, then I WILL have the chance to depart from my vacation and come home where the rest of my family and friends will be there waiting. I know that you know my heart, and that my intentions are good but my humanity causes me to struggle with this concept. I ask you to come into my heart and my soul and show me how to be more like you. I want to live and work, as well as laugh and create healthy debates all in the honor of you since you died to give me the chance to do this so and thereafter have complete harmony with those I love! Please accept my plead and help me to keep it even when I still have my doubts. Do not give up on me Jesus… you said you will never leave me or forsake me. This means you will not leave my side when things get bad. Things will be bad at times but you are ALWAYS there to get me through it and to offer me the peace in knowing that there is light no matter which way it comes out. Thank you Jesus for listening to me and for coming into my Heart and my Spirit! I look forward to the day I meet you and see my family again. Those who believe will not perish but have everlasting life. I am on vacation and will return home when I do. And when I do, I will return home because I have prayed this prayer. Thank you Jesus for this brilliant and loving gift. It is irreplaceable and it is not able to be measured up to in any way. AMEN!

Copyright 2010 Kristina DesJardins

 copyright

all rights reserved

www.ingramcontent.com/pod-product-compliance
Lightning Source LLC
Chambersburg PA
CBHW021934170526
45157CB00005B/2315